Leandro Galzerano
Roberto Rossiello
Eliane Morgado

Establishing pastures of Tifton 85 grass

Leandro Galzerano
Roberto Rossiello
Eliane Morgado

Establishing pastures of Tifton 85 grass

Speeding up processes

ScienciaScripts

Cover image: www.ingimage.com

This book is a translation from the original published under ISBN 978-613-9-65274-7.

Publisher:
Sciencia Scripts
is a trademark of
Dodo Books Indian Ocean Ltd. and OmniScriptum S.R.L publishing group

120 High Road, East Finchley, London, N2 9ED, United Kingdom
Str. Armeneasca 28/1, office 1, Chisinau MD-2012, Republic of Moldova, Europe
Printed at: see last page
ISBN: 978-620-7-85913-9

TABLE OF CONTENTS

CHAPTER 1	**3**
CHAPTER 2	**5**
CHAPTER 3	**10**
CHAPTER 4	**16**
CHAPTER 5	**30**

SUMMARY

This work was carried out with the aim of defining nitrogen fertilisation strategies in Tifton 85 pastures, based on monitoring dynamic changes in the structure of the forage canopy. The experiment was carried out in an area of Planossolo in the experimental field of the Federal Rural University of Rio de Janeiro, in the municipality of Seropédica, RJ, during the pasture establishment phase, between October 2006 and January 2007. The treatments consisted of four levels of N (0; 50; 100 and 150 kg N-urea.ha'^{1}), in a single application, arranged in 4 x 4 m plots, arranged in completely randomised blocks, with four replications. Together with the nitrogen fertiliser, the plots received 80 kg P.ha'^{1} and 150 kg K-ha^{1} . The soil was corrected with 1.0 tn.ha'^{1} of limestone. The variables studied were: frequency of plant occurrence; canopy height, leaf growth and leaf area index (LAI), interception of photosynthetically active radiation (PAR) and dry matter production and partitioning. The results obtained showed that the frequency of occurrence; height and RFA interception varied temporally according to a logistic pattern, the parameters of which were modified, to a greater or lesser degree, by the levels of N added. Similarly, at 87 days after planting, IAF and the total dry matter produced, but not its partitioning between leaves and stalks, responded to variations in the mineral N available in the soil. It was concluded that the application of increasing levels of N-urea during the establishment phase of the Tifton 85 grass was an efficient experimental strategy for inducing quantitative variations during the temporal development of the canopy structure.

Keywords: *Cynodon* spp. Frequency of occurrence. leaf area index. Photosynthetically active radiation. Leaf:stem ratio. Total dry matter.

CHAPTER 1

INTRODUCTION

In Brazil, from the 1960s onwards, there was a huge expansion in the area under pasture to meet domestic demand for animal products (BOMFIM DA SILVA, 2005). The end of the 20th century was marked by a need to re-evaluate attitudes and procedures in various sectors, due to economic stability (PEDREIRA et al., 2002). In this context, the primary sector was forced to direct its efforts towards technification, with the aim of increasing the efficiency of the production process. One of the sector's activities that has perhaps been hit hardest is livestock farming, which in Brazil is essentially based on the use of pastures.

In the coming years, the price of fossil fuels will rise, and grains will increasingly be used as a raw material for biofuel production, which in turn will increase the cost of this vegetable raw material. However, these changes will set limits for animal production systems based on the use of grains. From this perspective, the production of animal protein from ruminants kept under grazing becomes a guaranteed sustainable alternative (LENG, 2002).

Given these premises, obtaining satisfactory levels of animal productivity in pastures necessarily requires an understanding of the processes and management strategies of the soil-forage plant system.

Pasture and grazing management, together with the introduction and evaluation of new grass and legume cultivars, have long been priority targets for experimentation with tropical forage plants in Brazil (DA SILVA and NASCIMENTO

JÚNIOR, 2007). According to Sbrissia et al. (2007), there has been significant progress in recent years with regard to understanding the factors that condition forage production in tropical pastures. Basically, the change in research focus requires the use of experimental treatments focussed on stricter control of canopy structure, a central concept in the attempt to produce consistent information that can be reproduced in the country's different soil and climate conditions.

A frequently reported observation is that, a few years after they are established, pastures begin a process of decline in their productivity, evidenced by the invasion of unpalatable weeds, the appearance of bare areas and the crusting of the soil, with a consequent reduction in animal productivity rates. It is likely that this process is due to a combination of a series of negative factors, such as poor pasture formation, excessive animal stocking and failure to adopt fertilisation and soil conservation practices (MACEDO, 2005).

On the other hand, the exploitation of improved species capable of producing quantity and quality must be seen as fundamental, and it is in this scenario that forage plants of the *Cynodon* genus have stood out, as they have high production potential, good nutritional value and great flexibility of use (FAGUNDES et al., 1999).

Rocha et al. (2002) found that nitrogen fertilisation results in an increase in dry matter production and reported that one of the main problems in the productivity of tropical pastures is nitrogen deficiency, which results in a sharp drop in carrying capacity and animal gain.

In view of the above, this study was carried out with the aim of defining nitrogen fertilisation strategies in Tifton 85 pastures, based on monitoring dynamic changes in the structure of forage canopies. To this end, an experiment was set up in the municipality of Seropédica, RJ, during the rainy season, where quantitative variations in various attributes and canopy properties were induced by managing the availability of N in the soil during the grass's establishment phase. The hypothesis underlying this experimental strategy was that the application of increasing levels of N is an efficient tool for promoting quantitative variations in various attributes and canopy properties.

The objectives of this study were to assess the effects of nitrogen fertilisation on: the frequency of plant occurrence, canopy height, leaf growth and the development of the leaf area index, the interception of photosynthetically active radiation and the production and partitioning of dry matter.

CHAPTER 2

LITERATURE REVIEW

1.1 Plant species

Tifton 85 is a bermudagrass grass launched in 1992 by the Coastal Plain Experiment Station in Tifton, Georgia (BADE, 2000), and is an F1 hybrid from the cross between a strain introduced from southern Africa (PI 290884) and Tifton 68.

This cultivar is taller and has wider leaves and stalks than most commercially available bermudagrass. Its rhizomes are larger than those of Tifton 44, but in smaller quantities. Tifton 85 propagates very quickly by stolons, with more vigour than other varieties. For this reason, it is planted using green rhizomes or stolons, rather than dormant rhizomes.

In experiments carried out at the Coastal Plain Experiment Station, tifton 85 produced 26 per cent more forage and 11 per cent more digestibility than Coastal bermuda grass. In another study in Georgia, tifton 85 was compared to tifton 78 in a study with heifers under continuous grazing between April and October. The results showed a 40 % greater weight gain in favour of Tifton 85 (468 kg vs. 334 kg). These results show the productive superiority of tifton 85 over other older bermudagrass (BURTON, 2001; CHAMBLISS and DUNAVIN, 2003).

In Brazil, there is no precise record of where and how the *Cynodon* genus was introduced. The fact is that it may have been the initiative of producers and researchers to carry out behavioural assessments in Brazilian conditions (VILELA and ALVIN, 1998).

Tifton 85 is a C4 photosynthetic cycle forage (ATHAYDE et al., 2005) that adapts to both cold and hot climates; however, it has a very marked seasonality of mass production, almost completely ceasing to grow during the dry season. It is characterised as more cold-tolerant than Tifton 68 and is recommended for haymaking due to its leaf/stem ratio and palatability (REIS et al., 2005).

According to Nippert et al. (2007), in ecosystems with several limiting factors, the dominance of C4 grasses may depend on their ability to maximise, when available, the use of favourable conditions, and thus grow rapidly and persist during limiting periods, which can change during and between years.

According to Vilela (2005), there is a wide range of advantages to grasses of the *Cynodon* genus. Among them are their high productivity per area and good quality, associated with a high capacity to respond to fertilisation; great resistance to trampling; good capacity to adapt to different types of soil and climate, which gives them good resistance to wet soils and low temperatures. These points distinguish this genus from others that predominate in tropical conditions and justify its use as a promising alternative for producers seeking efficiency in dairy farming through sustained intensification of the activity. However, according to Lima and Vilela (2005), it is essential to look at factors such as the soil's fertility potential for forage production, so that the stocking rate can be set according to forage availability. In this context, it is important to note that the appropriate use of fertilisers increases forage yield and, consequently, greater grazing pressure can be adopted, with positive results for animal productivity.

1.2 Nitrogen

Nitrogen (N) is an essential component of amino acids and proteins, nucleic acids, hormones and chlorophyll, among the organic compounds essential to plant life (LAVRES JÚNIOR and MONTEIRO, 2003), appearing in the reduced form (NH2). Generally speaking, the ammonium cation ($NH4^+$), which comes from the deamination of decomposing organic matter (ammonification) or the addition of ammonia fertilisers, and the nitrate anion (NO3), which results from the biological oxidation (nitrification) of ammonium, are the common forms of N-mineral available to root systems. The total levels of NO3' and $NH4^+$ in the soil depend on the type of source and quantity of N added, as well as the balance between the processes of ammonification, nitrification, immobilisation and denitrification, which are modulated by environmental fluctuations and the pasture use system (VENDRAMINI et al., 2007). Nitrification plays a central role in the functioning of pastoral ecosystems, since as well as being directly involved in plant nitrogen nutrition, it regulates the magnitude of N losses through nitrate leaching and/or denitrification (FERNANDES and ROSSIELLO, 1995; ENGELS and MARSCHNER, 1995, LATA et al., 2004).

Of all the mineral nutrients, N is quantitatively the most important for plant growth (ENGELS and MARSCHNER, 1995) and is actively involved in promoting morphogenetic events, thus determining the parameters that define grass structure (CHAPMAN and LEMAIRE, 1993; CRUZ and BOVAL 2000; GASTAL and LEMAIRE, 2002). In soils deficient in N, plant growth and development are slow, tiller production is negatively affected and the protein content is deficient to meet the animal's requirements.

Numerous studies have evaluated the importance of N for pasture growth, both in temperate and tropical climates. For example, Sarmento et al. (2006), testing nutrients that limit the development of Tifton 85 grass in a Red-Yellow Argissolo, concluded that N was the most limiting nutrient for grass development.

There are two ways in which the plant obtains N: one internal, represented by the redistribution of N associated with leaf senescence, and the other external, through the deposition of faeces and urine, rainfall or the application of fertilisers (RUSSELLE, 1992), which can result in losses through leaching and gaseous emissions (EUCLIDES et al., 2007).

The rate of N acquisition by crops is highly variable during development and between years; however, under ample N availability in the soil, N accumulation in plants is highly related to the growth rate and biomass accumulation of the crop (GASTAL and LEMAIRE, 2002).

Marcelino et al. (2003), evaluating the effect of nitrogen fertilisation and water tensions on dry matter production and leaf area index of Tifton 85 grass in cerrado conditions, found that under favourable climatic conditions, an increase in N doses resulted in correlated increases in IAF. However, as Lugão et al. (2003) observed, it is necessary to know the appropriate dose of application, capable of economically maximising the productive potential of the pasture, at which point losses are minimised and the nutrient is used to its greatest efficiency.

With regard to the nutritional quality of *Cynodon* grasses, Rocha et al. (2002) point out that nitrogen fertilisation not only increases the volume of dry matter produced, but also the crude protein content, while promoting a proportional reduction in neutral detergent fibre.

Alvin et al. (2003) studied the effects of two doses of N:K combinations on four *Cynodon* spp. hybrids (African star grass, florona, florico and florakirk), finding a significant increase in the leaf:stem ratio of the florona and florakirk forage plants in response to fertiliser application. This increase in the leaf:stem ratio makes it easier for the animals to eat the leaves and improves the quality of the diet, resulting in better performance. Leaves are an essential part of the diet consumed by grazing animals. Carvalho et al. (2005) pointed out that animals kept on pastures, even when they have low percentages of blades and a high percentage of dead material, manage to have more than 80% of their diet made up of leaves.Alvim and Botrel (2001), studying doses of 100, 250 and 400 kg N.ha '.Alvim and Botrel (2001), studying doses of 100, 250 and 400 kg N.ha ' per year on Coastcross grass, concluded that a dose of 250 kg/ha.year of N maximised the stocking rate of the pasture and milk production per area, resulting in a higher economic return per unit area; the dose of 100 kg.ha^{1} , per year of N resulted in higher milk production per kilogram of N applied to the pasture.

1.3 Establishing Pastures

One of the most important requirements for successful pasture farming is to guarantee the pasture's longevity. In this context, good establishment is one of the most important aspects for achieving sustainability from pasture formation onwards (XAVIER et al., 2001; LIMA and VILELA, 2005).

According to Oliveira et al. (2006), a sufficient but not excessive amount of heat is a basic prerequisite for plant life and productivity where each physiological process is adjusted within a temperature range.

When establishing a pasture area, forage plants depend on the process of photosynthesis to produce the energy needed for their maintenance and growth. To do this, plants need to set up an efficient system for collecting incident radiation, usually based on specific combinations of leaf area index (LAI) and leaf angulation (ALEXANDRINO et al., 2004).Initial observations on the establishment of Tifton 85 grass on Florida farms showed rapid stolon growth compared to other *Cynodon* spp. hybrids (CHAMBLISS and DUNAVIN, 2003). On the other hand, considering that N also stimulates the speed of blade renewal (HOESCHL, 2005), it is inferred that the nutrient controls the spatial and temporal variations in IAF, with repercussions on the grass's light interception patterns.

CHAPTER 3

MATERIAL AND METHODS

3.1 Location of the Experimental Area

The experiment was conducted in an area belonging to the "Wilhelm Otto Neitz" Parasitology Experimental Station of the Veterinary Institute of the UFRRJ, Seropédica - Rio de Janeiro (22°45'53" S, 43°41'56"W and 33m altitude). The total period of the experiment was 87 days, from 3 November 2006 to 29 January 2007.

3.2 Edaphological characteristics

According to the current Brazilian Soil Classification System (EMBRAPA, 1999), the pedology of the experimental area is characterised as Planossolo Háplico. The chemical properties of the soil under study were characterised through sampling in the 0-20 cm layer carried out in July 2006. Routine analytical data was obtained from the UFRRJ Soil Fertility Laboratory (Tabelai).

Table 1. Results of the soil fertility analysis of the experimental area.

Professor	**Ca^{2+}**	**Mg^{2+}**	**K^{+}**	**H+Al**	**V**	**P(*)**	**M.O.(**)**	**PH**
(m)	(cmoVdm)^3	(cmof/dm)^3	(cmof/dm)^3	(cmof/dm)^3	(%)	(mg/dm)^3	(g/kg)	(water, 1:2.5)
0,0-0,2	1,8	1,0	0,079	0,8	78	14,5	1,69	5,4

(*) Available P (Mehlich-1).

(**) Organic C (Walkey-Black) x 1.72.

3.3Climate of the region

The region's climate is characterised as Aw, according to the Kõppen system, with two distinct seasons: a dry one, from April to September, and a rainy one, generally from October to March.

The agrometeorological data recorded during the experimental period for air temperature, rainfall and sunshine was obtained from the Agricultural Ecology Station km 47- Seropédica - INMET/PESAGRO-RIO, approximately 1.5 km from the experimental area.

3. 4Calage

In September 2006, lime was applied to ensure the soil's acidity was neutralised and base saturation was raised to 60-70%, the appropriate range for *Cynodon,* as indicated by Lima and Vilela (2005). The limestone was distributed evenly by hand at a dose of 1000 $kg.ha^{-1}$, followed by harrowing for incorporation at a depth of 25-30 cm.

3.5 Fertilising and planting

Planting took place on 3 November 2006, using rooted grass stolons, in furrows 1.0 m apart. Also at the time of planting, the plots received a basal application of P and K, at doses of 80 $kg.ha^{-1}$ and 150 $kg.ha^{-1}$ respectively, in order to eliminate differences in the availability of these elements.

The fertilisers used were: Super Simple Phosphate and Potassium Chloride as sources of Phosphorus and Potassium respectively.

3.6Treatments and Experimental Design

The treatments consisted of doses of N as follows: control treatment (no N added), 50 kg $N.ha^{-1}$, 100 kg $N.ha^{-1}$ and 150 kg $N.ha^{-1}$.

The source of N used was agricultural urea diluted in water, and the respective doses (treatments) were applied to the rows at planting time using plastic watering cans, as described by Magiero (2004).

The treatments were arranged in four blocks, where they were randomised. Each experimental unit (plot) had dimensions of 4 x 4 m ($16m^2$), with the central 2 x 2 m part being considered a useful area for sampling or phytomass collection. Each plot was separated from its neighbour by a 1.0 m external border, which served as a circulation area for collecting samples and cultivation.

The mathematical model used was:

$y_{ij} = m+t +b +e_{ijij}$

Where:

-y_{ij} *is* the observation for treatment i in block j;

-m *is* the overall average;

-t_i *is the* effect for treatment i, where i = 1 to 4;

-b_j is the effect of block j, where j = 1 to 4;

-and_{ij} *is* the experimental error associated with observation ytj.

3.7 Monitoring Experimental Variables

The sampling dates, calculated as days after planting (DAP), for each of the attributes studied can be seen in Table 2.

Table 2. Monitoring of the variables studied.

Variable	Days After Planting (DAP)					
Frequency	30	35	40	47	55	64
Canopy height	30	35	40	55	64	__(D
Rad interception. Photosynthetically. Active	33	44	69	86	-	-
Specific leaf area	76	-	-	-	-	-
Fodder mass	87	-	-	-	-	-
leaf area index	87	-	-	-	-	-

(1)-: not applicable.

3.8Variables studied

3.8.1Frequency of plant occurrence

Determining the frequency of occurrence of plants was used in this study as an indirect estimator of the speed of pasture establishment, quantifying the spatial distribution of aerial tillers. To do this, an intersection method was used, according to Ribeiro et al. (2007). A 4 m quadrant^2 (2m x 2m), with subdivisions separated by plastic wires every 0.2 m, totalling 81 intersection points, was placed on top of the grass, adjusted to the height of the canopy and always in the same position.

The absolute frequency of the grass was calculated using the expression (1):

$$\text{Absolute frequency} = \text{NP17NTP.100} \quad ....(1)$$

where: NPT, is the number of intersections touched by some element of vegetation and NTP is the total number of possible intersections within the quadrant.

3.8. 2Canopy height

It was determined using a ruler graduated in millimetres, in association with frequency determination. As a result of the unevenness of the grass, the "uncompressed" height method was used, taking the average height between the highest and lowest point under a transparency sheet placed on the surface of the forage canopy (DA SILVA and CUNHA, 2003).

3.8.3Interception of photosynthetically active radiation (RFA)

The photosynthetically active radiation incident on the canopy was measured using a specific *quantum* sensor (400 - 700 nm), model LI 190 SA (LICOR, Nebraska, USA) connected to a LI 250 portable radiometer (LI-COR, Nebraska, USA), as described by Nobel and Long (1988). On each sampling date (Table 2), measurements were taken of incident photosynthetic irradiance (μmol_{foton} m^{-2} s^{-1}), where the sensor was positioned above the canopy, and of transmitted photosynthetic irradiance, where the sensor was positioned at ground surface level. The device was always operated in "Average" mode, which provides an average value of eight readings after 15 seconds of intermittent measurement.

The sensor was moved randomly over the surface of the canopy so that several points were sampled. The fraction of incident photosynthetically active radiation (incident RFA) effectively intercepted by the canopy (RFA int) was calculated using the following expression (2):

$$\text{RFA\^{}W} = 100\text{x} \frac{(\text{RF Aincident}_{\text{at the top-}} \text{RF A}_{\text{transmitted to the ground}})}{\text{RFA}_{\text{incident on the top of the canopy}}} \quad ....(2)$$

3.8. 4Forage mass production and partitioning

At the end of the experimental period, the standing forage mass was cut from a 0.25 metre quadrant. After collection, the material was taken to the Bromatology Laboratory to fractionate the green mass after removing the weeds. The material was fractionated into leaf blades, stalks + sheaths and senescent material. In order to minimise dry mass losses due to the effects of dehydration of the green material, the cuts were made sequentially, collecting a single block (four plots) at a time. The fresh mass of each component was recorded on a scale to two decimal places immediately after separation. The material was then dried in a ventilated oven at 70° C until it was a constant mass and then weighed on an analytical balance. Based on these determinations, the biomass of the aerial part per surface unit was estimated, as well as its distribution between leaves (blades), stalks + sheaths and senescent material.

3.8.5 Leaf area index (LAI)

The leaf area was assessed using a planimetric procedure to determine the Specific Leaf Area (SFA) for each plot.

The EFA was determined using 10 discs of leaf blades (1 disc/leaf) obtained with a paper punch with a known area (0.3002 cm /disc), from pre-selected leaves representative of each plot and visually classified as small, medium or large. Given their small size, the discs were handled carefully and dried immediately in a ventilated oven (70° C for 48 hours).

After drying, the weight of the discs was recorded on a precision scale. The EFA data was expressed in units of $cm^2 .g^{-1}$, and used to determine the Leaf Area Index (LAI).

The IAF was estimated by multiplying the AFE by the mass of leaf dry matter (MSF) collected from a known surface area. To do this, the mass of leaf dry matter from a 0.25m quadrant was determined[2] , according to the protocol in 3.8.4. The IAF value was then calculated as the product (expression 3):

$$IAF = g.\ MSF/2500\ cm^2 \times AFE\ (cm^2\ leaf/g\ MSF)\ldots\ldots\ldots\ldots(3)$$

Recently, Sbrissia and Da Silva (2008) validated a similar destructive procedure for determining IAF in marandu grass pastures under continuous stocking.

3.9 Statistical Analyses

The measured parameters were subjected to standard tests for normality of distribution and then processed for analysis of variance using the SAEG programme, using the sub-procedure of repeated measures over time. This analysis of variance consisted of the initial separation of significance levels for N, according to the "F" test. However, as the two main independent variables involved were quantitative in nature (N doses and time, the latter in the form of Days After Planting, DAP), the final analysis was conducted according to the principles of regression analysis.

CHAPTER 4

RESULTS AND DISCUSSION

It was considered appropriate to begin this section with a brief analytical description of the environmental conditions in force during the experimental period, relating to rainfall, sunshine and air temperature.

4.1 Environmental Conditions During the Trial Period

The spatial distribution of solar radiation, through energy balances, is a preponderant factor in the occurrence of physical and biological phenomena in the various regions of the globe, but it is the water factor that significantly affects plant growth and development (ORTOLANI and CAMARGO, 1987).Table 3 shows the rainfall values, grouped into decennia, during the experiment, which totalled 594, Imm. It can be seen that although the rainfall on 11/10 - 10/11 2006 was low (10.4 mm), there was satisfactory rainfall during all the intervals of the experimental period.

Table 3. Distribution of daily precipitation in the decennia and total during the experimental period.

Period	**Rainfall (mm) on each day of the period**										**Total (mm)**
1-lO/Out	11,8	00	4,8	00	00	8,0	**79,6**	2,6	2.6	00 -	109,4
11-20/Oct	00	00	00	00	00	00	00	3.2	5,4	**12,6** -	21,2
21-31/Oct	**5,0**	00	4,6	00	00	00	0,8	00	00	00 00	10,4
Month of Oct											**141,0**
1-10/Nov	00	0,4	3,2	00	00	00	0,2	4,8	1,2	**8,0**	- 17,8
11-20/Nov	6,6	5,2	0,4	13.4	**14,3**	0.2	0,9	00	00	3,6	- 44,2
21-30/Nov	3,2	00	00	00	00	00	1,4	24	**24,7**	19	- 72,3
Month of Nov											**134,3**
1-10/Dec	10,6	00	00	00	00	00	**19,4**	0,9	00	2,8	- 33,7
11-20/Dec	00	8,2	0,8	1,4	3,6	**28,1**	00	00	00	00	- 42,1
21-31/Dec	3,2	**16,2**	00	12,5	3,8	3,2	0,8	1,2	1,2	15,6	11 68,7
Month of Dec											**144,5**
1-10/Jan	16,8	6,0	19,6	**21,6**	1,8	0,4	00	3,0	4,8	8,4	- 82,4
11-20/Jan	00	00	**35,4**	4,4	00	3,8	10	6,4	0,2	00	- 60,1
21-29/Jan	00	7,4	**11,8**	0,2	00	5,0	2,6	00	4,8		- 31,8
Month of Jan											**174,3**
Period Exp.											**594,1**

Source: Agricultural Ecology Station km 47- Seropédica - INMET/PESAGRO-RIO.

The data in Table 3 shows the unevenness of rainfall within the decennia, a fact that is considered normal, but which shows pronounced differences in the maximum intensity of daily rainfall. Thus, during the month of October, from the 1st to the 10th[2] day, 79.6 mm of rain was recorded on a single day (07/10), which corresponds to the normal rainfall (76.1 mm) for the month of October (MARTORANO et al., 1997). However, taking into account the days without rain and the rainy days within the decennia (Table 4) shows that, with the exception of the 1st decennium of December, where it is possible that the rain that fell did not fully compensate for the daily evapotranspiration levels, the rest of the period did not have any dry periods of considerable amplitude.

Table 4. Days of rain and no rain, maximum precipitation in 24 hours and maximum duration of the dry period during the experimental period.

Period	Days without rain in period	Rainy days in the period	Max. duration dry period (days)
O-lO/Out	4	6	2
11-20/Oct	7	3	7
21-31/Oct	8	3	4
Month of Oct	**19**	**12**	**7**
1-10/Nov	4	6	3
11-20/Nov	2	8	2
21-30/Nov	5	5	5
Month of Nov	**11**	**19**	**5**
1-10/Dec	6	4	5
11-20/Dec	5	5	4
21-31/Dec	1	10	1
Month of Dec	**12**	**19**	**5**
1-10/Jan	1	9	1
11-20/Jan	4	6	2
21-29/Jan	3	6	1
Month of Jan	**8**	**21**	**2**

Source: Agricultural Ecology Station km 47- Seropédica - INMET/PESAGRO-RIO.

The average daily insolation values can be seen in Figure 1, which are inversely related to those corresponding to rainfall, i.e. during periods with rain, insolation was low. In the intervals in which there were more days without rain, the insolation recorded was high, which is an indication of high atmospheric transparency, i.e. clear skies without cloudiness, favouring a high incidence of global solar radiation.

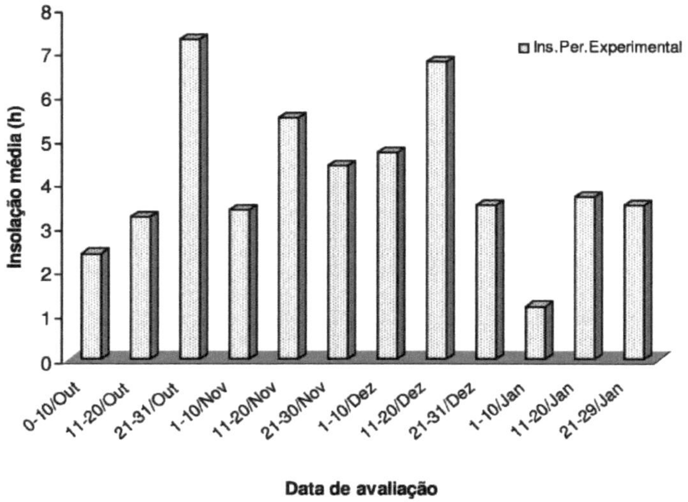

Figura 1. Distribution of sunshine (h) during the experimental period.

Table 5 shows the minimum, average and maximum air temperatures during the experimental period. Temperatures fluctuated little with minimum values above 19°C (except for the period 21-31/Oct with 17°C) and maximum values above 27°C. Considering the nature of the metabolism of tifton, a C4 species (ATHAYDE et al., 2005), it can be inferred that the temperatures in force were not limiting to the expression of the photosynthetic productivity of the pasture.

Table 5: Average monthly values for maximum temperature (TMAX), minimum temperature (TMIN), average temperature (TMÉDIA) and hours of sunshine during the experimental period.

Date	TMAX (°C)	TMIN (°C)	AVERAGE (°C)	Insolation (h)
O-lO/Out	27	19	22	23
11-20/Oct	28	20	23	32
21-31/Oct	28	17	22	80
1-10/Nov	27	19	23	34
11-20/Nov	29	19	23	55
21-30/Nov	31	22	25	44
1-10/Dec	30	20	24	47
11-20/Dec	32	22	26	68
21-31/Dec	31	22	26	39
1-10/Jan	28	22	25	12
11-20/Jan	29	22	25	37

21-29/Jan	31	22	26	32

Source: Agricultural Ecology Station km 47- Seropédica - INMET/PESAGRO-RIO.

4.2Effects of N on Pasture Establishment

Table 6 shows a general summary of the results of the analyses of variance (based on the raw primary data, since there was no need for transformations) expressed through the statistical probabilities associated with the "F" test.

Table 6. Statistical probability levels (F test) for the analyses of variance of the experimental variables studied.

Variable	Doses of N	DAP	Treatment x DAP
FREQ	0,04467	0,0000	0,1236
ALT	0,0324	0,0000	n.s(1)
IAF	0,03004	—(2)	-
IRFA	0,0006	0,0000	0,056
MSTPA	0,00493	-	-
MSF	0,01069	-	-
MSC	0,02406	-	-

(1) n.s.: non-significant (p> 0.05); (2) --------: not applicable.

DAP: Days after planting; FREQ = frequency of plant occurrence; ALT = canopy height; IAF = leaf area index; IRFA = interception of photosynthetically active radiation; MSTPA = total dry mass of the aerial part; MSF = dry mass of leaves; MSC = dry mass of stalks.

4.3Frequency of Plant Occurrence

The analysis of variance showed significant effects ($p < 0.05$) for N doses and age of growth, but not for their interaction. Based on this result, a regression analysis was carried out and a Boltzmann-sigmoidal function was selected, which fitted the data fairly accurately, adequately explaining the shape of the spatial variation, which tends to reproduce the well-known logistic pattern of dry matter accumulation (PARSONS et al., 2001). To improve the accuracy of the estimate, an additional Frequency value was introduced, assuming zero frequency at the time of planting, i.e. for DBH = 0, frequency = 0.

It is interesting to note that the area where the quadrant was initially placed was demarcated with wooden stakes. These stakes remained fixed and the quadrant was inserted between them in successive evaluations to ensure that the samples were always taken over the same area. This precaution allows us to assume that the evaluations do in fact reflect land cover speeds caused by the differential effects of N on stolon growth rates.

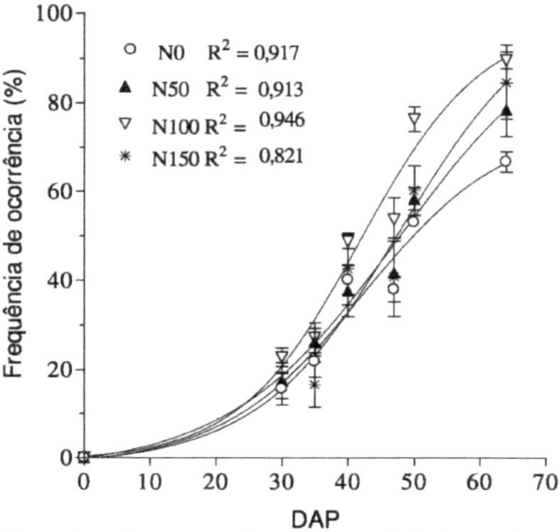

Figura 2. Frequency of occurrence of plants as a function of nitrogen doses and days after planting.

On the day of the final assessment (64 D AP) the average frequency values were 66.7, 78.4, 89.51 and 84.6 per cent, for the levels of 0, 50, 100 and 150 kg N.ha'[1] , respectively. The analysis of variance focused only on this last collection detected a significant difference ($p<0.05$) between the value of lOOkgN applied and the control (Figure 2).Although the dry mass data will be considered later, it is interesting to note that in the course of comparative analyses between the various canopy attributes, a significant correlation was obtained between the frequency data and the corresponding aerial part dry mass values (MSTPA) (r (Pearson) = 0.679; $\alpha < 0.05$; Figure 3). The form of this association is comparable to that obtained by Oliveira (2008) who cultivated Tifton 85 in a parallel experiment, using N levels between 0 and 600 kg N.ha^{-1} .

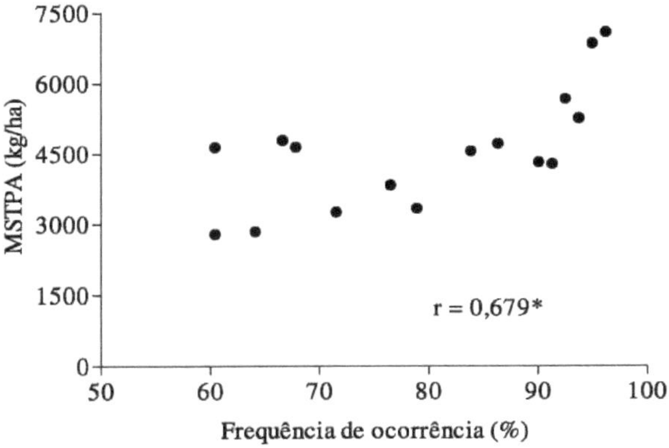

Figura 3. Relationship between the frequency of occurrence of Tifton 85 plants and the total dry mass of the aerial part (MSTPA).

4.4 Canopy height

For this canopy attribute, the effects of N doses, although significant, were less consistent in terms of discriminating between treatments. As with frequency, the effects of N and sampling dates were independent. The data was analysed using the same regression procedure applied to the frequency data, and the results are shown in Figure 4.

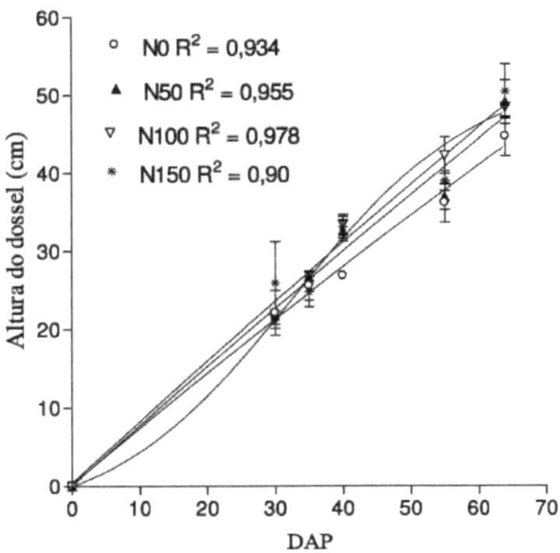

Figura 4. Canopy height as a function of nitrogen doses and days after planting.

The average heights, while dependent on the level of N, ranged from 31.0 cm (control) to 34.6

cm with 150 kg $N.ha^{-1}$. On the other hand, they varied linearly with the age of the pasture, from 27 cm (at 30 DAP) to 48 cm (at 64 DAP).

According to Brougham (1958), quoted by Pedreira et al. (2001), the maximum rate of net forage accumulation would occur when the pasture reached an interception level of 95% of the incident light. Although the level of photosynthetic irradiance interception will be considered later (item 4.6), it can be noted here that there is a relationship between canopy height and RFA interception (Figure 5).

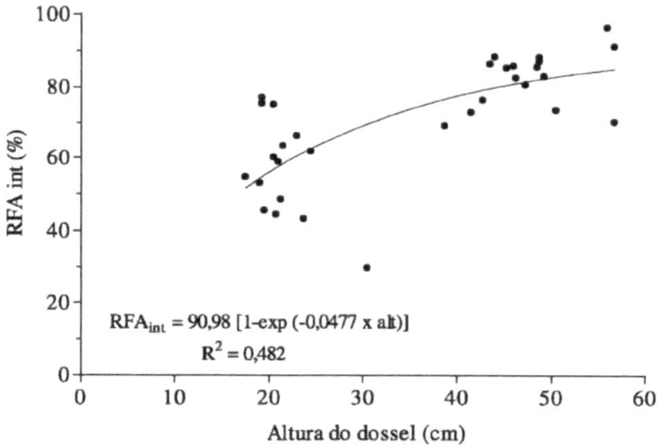

Figura 5. Relationship between canopy height and interception of photosynthetically active radiation.

The equation predicts a maximum interception of approximately 91 per cent, which does not deviate from the values of intercepted RFA actually recorded at 87 DAP (item 4.6).

4.5 leaf area index

The leaf area index (LAI) expresses the relationship between the leaf area of a canopy and the area of ground covered by its vertical projection (NOBEL and LONG, 1988). As such, it is the main factor influencing light interception and, consequently, dry matter accumulation rates, as well as the dynamics of regrowth, which is partly dependent on the residual IAF after cutting or grazing

(CARVALHO et al., 2007).

The IAF values obtained in this experiment, determined 87 days after planting, are shown in Figure 5. The doses of N caused a significant increase in the value of this morphophysiological attribute ($p = 0.03$, Table 6), in the order of 52 % for the highest dose of application (5.93 ± 1.73) compared to the unfertilised control (3.92 ± 1.08). The figure also shows that a quadratic function described the effect of N with an R = 0.32. This low value for the coefficient of determination can be explained by the high sample variability of this canopy variable, evidenced by the magnitude of the standard deviations from the mean (Figure 6).

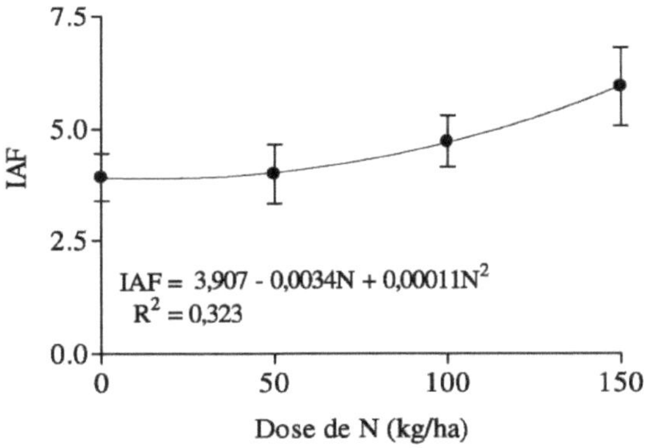

Figura 6. Effects of nitrogen application at planting on leaf area index.

In forage grasses, when growing freely, maintaining high IAF values is a necessity due to the angle of leaf insertion (NOBEL and LONG, 1988). Thus, the greater the erectophilous tendency in a given canopy, the greater the IAF value needed to achieve 95 % interception of the incident solar radiation (PEDREIRA et al., 2001). The IAF value associated with this condition is called "critical" (BROUGHAM,1958, cited by PEDREIRA et al., 2001), Fagundes et al. (1999) evaluated the effect of grazing intensities on IAF and light interception in pastures of three *Cynodon* spp. cultivars (coastcross, florakirk and tifton 85). In general terms, pastures kept higher showed higher values of IAF, *k* (the extinction coefficient) and light interception, which can be interpreted as a product of the larger, more horizontal leaves in these pastures, thus reducing the amount of light propagated along the canopy profile. On the other hand, the increase in IAF generates competition between plants, resulting in a change in the spectral composition of the light that reaches the leaves closest to the ground, with a reduced amount of blue light and a low far-red ratio (DEREGIBUS et al., 1983). This causes most of the axillary buds to remain dormant, i.e. there is a reduction or even a halt in the emission of new tillers (FRANK and HOLFMAN, 1994), which partly explains the stabilisation in canopy leaf area with age.

4.6 Interception of Photosynthetically Active Radiation

As previously mentioned, closely linked to IAF is the interception of photosynthetically active radiation (RFA_{Int} , 400-700 nm), a key biophysical function within the plants' trophic programme (NABINGER and PONTES, 2001). Based on the work of Monsi and Saeki (1953, cited by SBRISSIA and DA SILVA, 2001), this relationship is given by the expression (4):

$$RFA_{fa}\,t = [\ 1 - \exp\ (-k\ IAF)] \text{ .... } (4)$$

where k represents the light interception or extinction coefficient. As the equation indicates, the model predicts that the fraction of light intercepted declines exponentially as a function of the IAF (or canopy depth). This trend has already been shown previously (Figure 5).

In this study, the analysis of variance of the data relating to RFA interception indicated highly significant effects ($p< 0.006$) for N doses and DAP, but not for their interaction ($P = 0.056$, Table 6). Again, the non-linear regression procedure was used, where the sigmoidal function explained most of the variability associated with the effects of N and DAP, with R values of 0.94 or higher (Figure 7). The effects of N seemed to manifest themselves early on, although the lack of experimental data before 33 DAP does not allow a conclusive judgement to be made. At this time, however, there was a significant contrast between the interception of the control treatment and those that received N, which recorded higher values of percentage interception of RFA. As the height and IAF increased over the days, the canopy made better use of the incident solar radiation. Thus, between 33 DAP and 86 DAP, RFA interception, regardless of available N, increased by 40 % (from 44 % to 83.9 %).

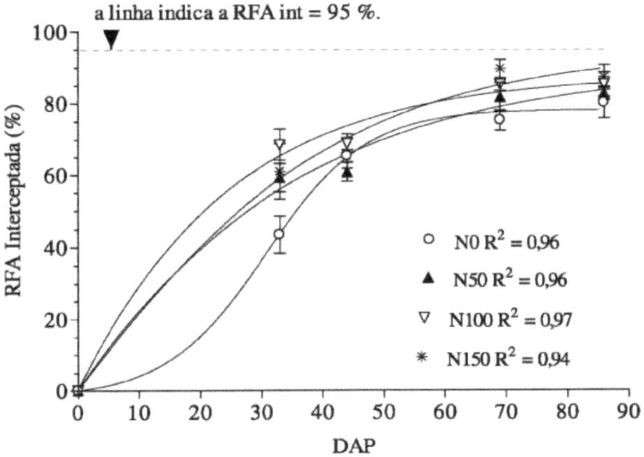

Figura 7. Variation in photosynthetically active radiation intercepted by Tifton 85 grass, as a function of days after planting (DAP).

Focusing on the last sampling date (Figure 7), it can be seen that the percentage range of differences in interception between the canopies corresponding to the highest application dose and those of the control treatment was only 7% (87.3 vs. 80.1%). This difference is much smaller than that corresponding to the IAF values of these treatments on the same date, of around 52%, as already mentioned (Figure 6).

According to Parsons (1988), light interception depends not only on the pasture's IAF, but also on the optical properties of the leaves as well as their spatial arrangement in the pasture.

canopy (this last factor is responsible for the value of *k* in equation 4). It is therefore possible that, in addition to differences in IAF and height between the canopies, nitrogen fertilisation induced changes in the leaf architecture of the canopy, since a community of forage plants in pastures can optimise IAF in different ways (RODRIGUES et al., 2006a). For this reason, the values of the interception coefficient *(k)* associated with this sampling date were estimated. Figure 7 shows the results obtained. The values were 0.445 ± 0.13; 0.479 ± 0.14; 0.432 ± 0.09 and 0.37 ± 0.03 for the levels of 0, 50, 100 and 150 $kg.ha^{-1}$ of N, respectively. These differences were not significant ($p>0.005$), although a quadratic pattern of variation was noticeable as the dose applied increased. On the other hand, the *k* values themselves are within those expected for grass canopies in general (NOBEL and LONG, 1988). This apparent failure to explain differences in RFA interception between canopies, based on the variables contained in equation 4, indicates the need to refine the canopy study methodology, introducing assessments of the stratified distribution of IAF within the canopy volume, i.e. considering its height component.

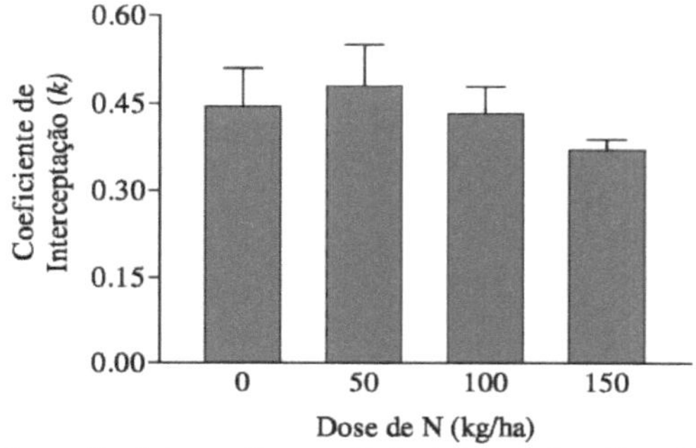

Figura 8. Interception coefficient *(k)* as a function of nitrogen doses.

4. **1Dry mass** ***production***

The analysis of variance revealed that both SBM and SBM production were altered only with the application of the highest dose of N. Therefore, in terms of forage mass production, only the contrast between the highest dose and the control was significant.

4. 8Total Dry Mass Production of the Aerial Part

Regression analysis showed that the quadratic fit model (Figure 9) passes through a minimum at N= 17 $kg.ha^{-1}$, which, however, has no biological significance. From this minimum, the quantitative increase in N produced the increases predicted by the model, up to the highest dose of nitrogen applied, where the MSPA yield obtained represented an increase of 51 % compared to the treatment that did not receive nitrogen fertilisation. It should be borne in mind that this value was obtained after a growth period of 87 days. Although the efficiency of use of the fertiliser applied (38.72 kg MSTPA/kg N-urea applied) is not particularly high, the result obtained is comparable to similar experiments. Although extrapolations should be made with caution, the work of Rocha et al. (2002) can be used as a comparison. These authors studied the responses of coastcross *(Cynodon* *dactylon (*L.) Pers x *(Cynodon nlemfuensis* Vanderyst), tifton 68 (*Cynodon* spp) and tifton 85 to the application of 0, 100, 200 and 400 kg.ha^{-1} , in an experiment set up in October 1997 in Lavras, MG. With regard to the production of tifton 85, the results obtained after three cuts (every 42 days) were 7.89 and 9.10 Lha^{-1} for applications of 100 and 200 kg N.ha^{-1} , respectively. Taking the average of these values (the equivalent of 150 kg N.ha^{-1}) and extrapolating the yield to 87 days gives a value of 6,000 kg.ha^{-1} , close to that of this experiment, regardless of differences in furrow spacing (0.5m in the work by ROCHA et al., 2002).

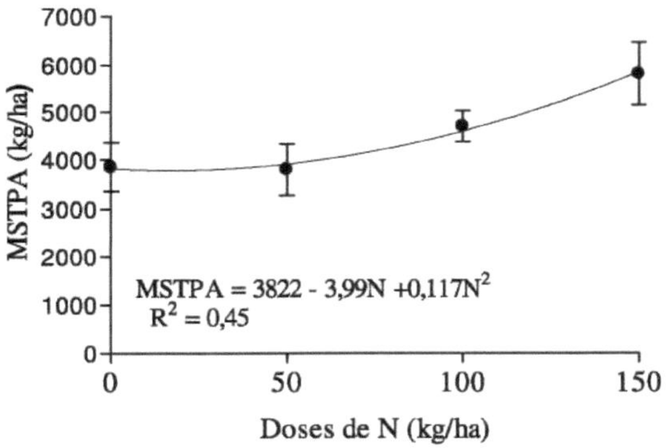

Figura 9. Effects of nitrogen application at planting time on total dry mass production.

4.9 Leaf Dry Mass Production

The values found for leaf dry mass production ranged from 1700 kg.ha^{-1} in the control treatment to 2458 kg.ha^{-1} at the dose of 150 kg.ha^{-1} of N (Figure 10), the only contrast to show a statistically significant difference ($p < 0.05$). Regression analysis showed that the quadratic model (Figure 10) has a minimum at N= 23 kg.ha^{-1} , which has no statistical or biological significance. From then on, the quantitative increase in N produces the increases predicted in the model, up to the highest dose of nitrogen applied, where the SPS yield obtained represented an increase of 44.8 % compared to the treatment that did not receive nitrogen fertilisation.

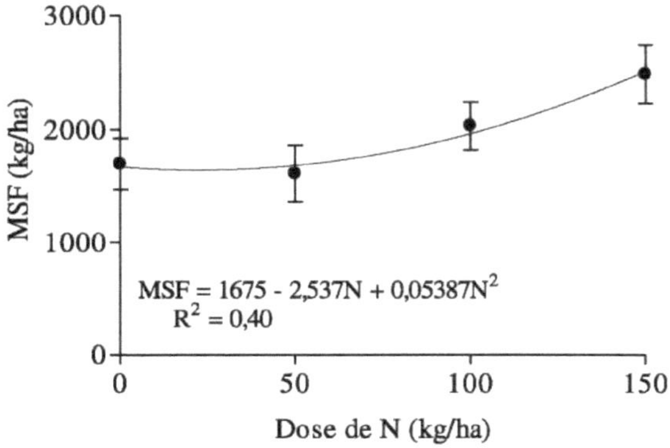

Figura 10. Effects of nitrogen (urea) application at planting time on dry mass production of leaves.

There is a wide range of variability in the literature for this important attribute related to forage consumption, which is why no specific comparison was made. What the data seems to suggest is that within the prevailing soil and climate conditions, there was some compensation in terms of mass allocation between leaves and stalks, probably because N affects both the population density of tillers (RODRIGUES et al., 2006; PATÊS et al., 2007) and the number of leaves per tiller (ALEXANDRINO et al., 2004).

4.10 Stalk Dry Mass Production

Figure 11 shows the regression between the doses of N applied and the component dry mass of the stalks (DSM), showing the quadratic model already seen for MST and MSF. At the highest level of application, the accumulation of dry mass in the stalks represented an increase of 55.8 % in relation to the treatment that did not receive nitrogen fertilisation. As already mentioned, the intermediate doses of N did not differ statistically ($p<0.05$) from the control treatment.

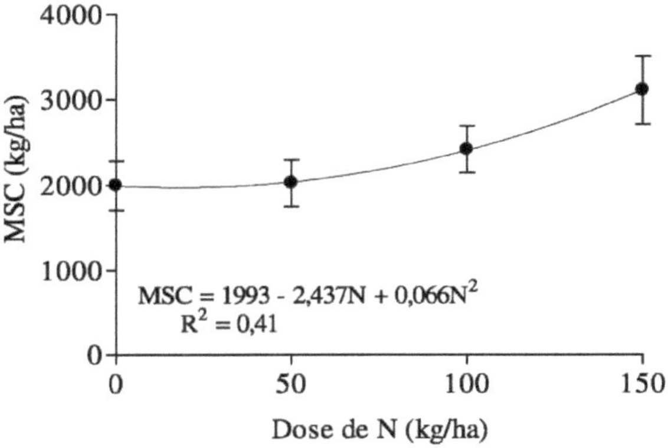

Figura 11. Effects of nitrogen (urea) application at planting time on dry mass production of stalks.

Managing forage plants with a view to having more green leaves or fewer dead leaves in the pasture favours the nutritional value of the forage, since leaves have a better chemical composition than stalks (RODRIGUES et al., 2006b). However, in this case, there was no pasture management to interfere with the partitioning of dry matter in favour of the leaves. The doses of N applied did not manage to alter the ratio shown by the control (Figure 12), as the differences did not reach statistical significance (p>0.05).

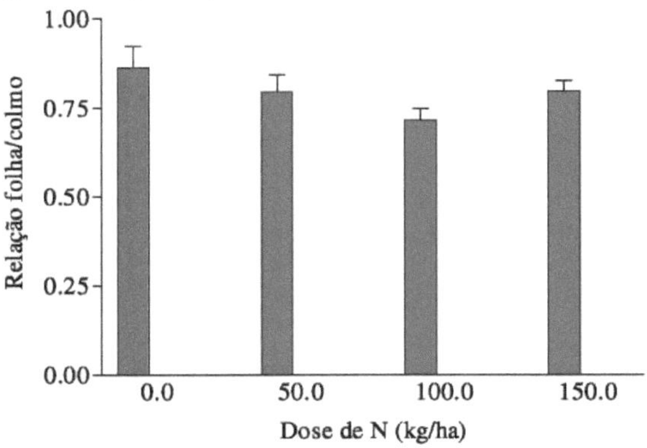

Figura 12. Leaf/stem ratio of Tifton 85 grass under doses of nitrogen fertiliser.

CHAPTER 5

CONCLUSIONS

The application of increasing levels of N-urea during the establishment phase of the Tifton 85 grass was an efficient tool for inducing quantitative variations during the temporal development of the canopy structure. Specifically, increasing the availability of N in the soil increased the speed of pasture establishment; it promoted leaf expansion and consequently the leaf area index and the interception of the associated photosynthetically active radiation. Thus, there were variations in the levels of dry matter accumulation in the aerial part, without altering its partitioning between leaves and stalks.

BIBLIOGRAPHICAL REFERENCES

ALEXANDRINO, E.; NASCIMENTO JÚNIOR, D.; MOSQUIM, P.R.; REGAZZI, A.J. Morphogenic and structural characteristics in the regrowth of *Brachiaria brizantha* cv. marandu submitted to three doses of nitrogen. **Revista Brasileira de Zootecnia,** v.33, n.6, p.1372- 1379, 2004.

ALVIM, J.M.; BOTREL, M.A.; REZENDE, H.; XAVIER, D.F. Evaluation under grazing of the forage potential of grasses of the genus *Cynodon* under doses of nitrogen and potassium. **Revista Brasileira de Zootecnia,** v. 32, n.2, p.47-54, 2003.

ALVIM, M.J.; BOTREL, M.A. Effect of nitrogen doses on milk production of cows on coast-cross pasture. **Pesquisa Agropecuária Brasileira,** v. 36, n. 3, p. 577-583, mar. 2001.

ATHAYDE, A. A. R.; CARVALHO, R. C. R,; MEDEIROS, L.T.; VALERIANO, A.R.; ROCHA, G.P. Grasses of the genus *Cynodon* - recent cultivars in Brazil. **Technical Bulletin** - n.^{e} 73, p. 1-14, Lavras, MG. 2005.

BADE, D.H. Bermudagrass varieties - Tifton 85, Jiggs, World Feeder. 55th Southern Pasture and Forage Crop Improvement Conference. **Proceedings...** Raleigh, 2000, p. 12-14.

BOMFIM DA SILVA, E.M. **Nitrogen and sulphur in the recovery of degraded Brachiaria grass pastures in Quartzarenic Neosol with significant organic matter.** 2005. 124 p. Thesis (Doctorate in Agronomy - Soils and Plant Nutrition) - Escola Superior de Agricultura "Luiz de Queiroz", Piracicaba, 125p., 2005.

BURTON, G.W. Tifton 85 bermudagrass-Early history of its creation, selection and evaluation. **Crop Science,** v. 41, p.5-6, 2001.

CARAVALHO, P.C. DE F.; GENRO, T.C.M.; GONÇALVES, E.N.; BAUMONT, R.A. Pasture structure as a management concept: effects on consumption and productivity. In: Reis, R. A. BERNARDES, T.F.; SIQUEIRA, G.R.; MOREIRA, A.L. (Ed.). **Volumes in Ruminant Production,** Jaboticabal, Ed. Funep. 2005, p. 107-124.

CARVALHO, C.A.B.; ROSSIELLO, R.O.P.; PACIULLO, D.S.C.; SBRISSIA, A.F.; DERESZ, F. Classes of tillers in the composition of leaf area index in elephant grass pastures. **Pesquisa Agropecuária Brasileira,** v.42, n.4, p.557-563, 2007.

CHAMBLISS, C.G.; DUNAVIN, L.S. **Tifton 85 bermudagrass.** 2003. Available at: <http://edis.ifas.ufl.edu>. Accessed on: 25/08/2007.

CHAPMAM, D.F.; LEMAIRE, G. Morphogenetic and structural determinants of plant regrowth after defoliation. In: BAKERS, M.J. (Ed.). **Grasslands for our world.** Wllington: SIR Publishing, 3, p.55-56,1993.

CRUZ, P.; BOVAL, M. Effect of nitrogen on some morphogenetic traits of temperate and tropical perennial forage grasses. In: LEMAIRE, G.; HODGSON, J.; MORAES, A.; NABINGER, C.; CARVALHO, P.C.F. (Ed.). **Grassland Ecophysiology and Grazing Ecology.** New York: CABI Publishing, USA, 2000, p. 151-168.

DA SILVA, S.C.; CUNHA, W.F. Indirect methods for estimating forage mass in pastures of Cynodon spp. **Pesquisa Agropecuária Brasileira,** v. 38, n. 8, p. 981-989, 2003.

DA SILVA, S.C.; NASCIMENTO JÚNIOR, D.; Advances in research with tropical forage plants in pastures: morphophysiological characteristics and grazing management. **Revista Brasileira de Zootecnia,** v. 36, special supplement, p.121-138,2007.

DEREGIBUS, V.A., SANCHEZ, R.A., CASAL, J.J. Effects of light quality on tiller production in Lolium spp. **Plant Physiology,** v.72, p.900-912, 1983.

EMBRAPA. **Brazilian Soil Classification System.** Brazil: 1999.412p.

ENGELS, C.; MARSCHNER, H. Plant uptake and utilisation of nitrogen. In: P.E, Bacon (Ed). **Nitrogen fertilisation in the environment.** New York: Mareei Dekker, Inc. 1995. p. 41- 81.

EUCLIDES, V.P.B.; COSTA, F.P.; MACEDO, M.C.M.; FLORES, R. Biological and economic efficiency of Tanzania grass pasture fertilised with nitrogen in late summer. **Pesquisa Agropecuária Brasileira,** v. 42, n.9, p. 1345-1355,2007.

FAGUNDES, J.L.; DA SILVA, S.C.; PEDREIRA, C.G.S.; SBRISSIA, A.F.; CARNEVALLI, R.A.; CARVALHO, C.A.B.; PINTO, L.F.M. leaf area index, light interception and forage accumulation in *Cynodon* spp. pastures under different grazing intensities. **Scientia Agrícola,** v.56, n.4, p.l 141-1150,1999.

FERNANDES, M. S., ROSSIELLO, R. O. P. Mineral nitrogen in plant physiology and plant nutrition. **Critical Reviews in Plant Sciences,** v.14, p.l 11-148,1995.

FRANK, A.B., HOFMAN, L. Light quality and stem numbers in cool season forage grasses. **Crop Science,** v.34, p.468-473,1994.

GASTAL, F.; LEMAIRE, G. N uptake and distribution in crops: An agronomical and ecophysiological perspective. **Journal of Experimental Botany. V.** 53, n.370, p. 789-799, 2002.

HOESCHL, A.R. **Pasture production of tanzânia grass fertilised with different doses of nitrogen under grazing.** Curitiba, PR, 2005. 72 p. Dissertation (Master's Degree in Agronomy) Federal University of Paraná.

LATA, J.C.; DEGRANGE, V.; RAYNAUD, X.; MARON, P. A.; LENSI, R.; ABBADIE, L.Grass populations control nitrification in savanna soils. **Functional Ecology.** v. 18, p. 605- 611,2004.

LAVRES JÚNIOR, J.; MONTEIRO, F.A. Tillering, leaf area and root system of mombasa grass submitted to combinations of nitrogen and potassium doses. **Revista Brasileira de Zootecnia,** v.32, n. 5, p. 1068-1075, 2003.

LENG, R.A. Future direction of animal protein production in a fossil fuel hungry world. **Livestock Research for Rural Development,** v. 14, n. 5, 2002.

LIMA, J.A.; VILELA, D. Formation and management of *Cynodon* pastures. In: VILELA, D.; RESENDE, J.C.; LIMA, J (Ed.). ***Cynodon,* forage crops that are revolutionising Brazilian livestock farming.** 1. ed. Juiz de Fora: Ed. Embrapa Gado de Leite, Brazil, 2005, p. 11-32.

LUGÃO, S.M.B.; RODRIGUES, L.R.A.; ABRAHÃO, J.J.S.; MALHEIROS, E.B.; MORAIS, A. Forage accumulation and nitrogen utilisation efficiency in *Panicum maximum* Jacq. pastures fertilised with nitrogen. **Acta Scientiarum. Animal Sciences,** v. 25, n. 2, p. 371-379, 2003.

MACEDO, M.C.M. Pastures in the Cerrados ecosystem: evolution of research for sustainable development. In: REUNIÃO ANUAL DA SOCIEDADE BRASILEIRA DE ZOOTECNIA, 42., 2005, Goiânia. **Proceedings..............................** Goiânia: SBZ/UFG, 2005. p.56-84

MAGIERO, J.Q. **Management strategies for nitrogen and potassium fertilisation in *Brachiaria humidicola* pasture, established on a planosol in the lowlands of Rio de Janeiro, during the rainy season.** Seropédica, RJ, 2004. 60p. Dissertation (Master's Degree in Zootechnics). Federal Rural University of Rio de Janeiro.

MARCELINO, K.R.A,; VILELA,L.; LEITE, G.G.; GUERRA, A. F.; DIOGO, J.M.S. Management of nitrogen fertilisation and water tensions on dry matter production and leaf area index of tifton 85 grown in the cerrado. **Revista Brasileira de Zootecnia,** v.32, n.2, p.268-275, 2003.

MARTORANO, L. G.; MATOS, R. M. B.; SILVA, E. M. R. Climatic availability for banana cultivation in the Seropédica / Itaguaí-RJ region. In: BRAZILIAN CONGRESS OF AGROMETEOROLOGY, X. Piracicaba: SBA, 1997. **Proceedings...** p. 301-303.

NABINGER, C.; PONTES, L.S. Morphogenesis of forage plants and pasture structure. In: MATTOS, W.R.S. et al. (Ed.). **Animal production as seen by Brazilians.** Piracicaba, SP: Ed. FEALQ, Brazil, 2001, p. 755 - 771.

NIPPERT, J.B.; FAY, P.A.; KNAPPA, A.K. Photosyntetic traits in C3 and C4 grassland species in mesocosm and field environments. **Environmental and Experimental Botany. V.** 60, p.412-420, 2007.

NOBEL, P.S. ; LONG, S.P. Dose structure and light interception. In: Coombs, J., Hall, D.O., Long, S.P., Scurlock, J.M. (Ed). **Techniques in Photosynthesis and Bioproductivity.** Colegio de Postgraduados, Chapingo, Mexico, 1988. p. 34-41.

OLIVEIRA, A.P.P. **Crescimento e composição químico-bromatológica do capim-tifton 85 *(Cynodon* spp.) sob doses de nitrogénio.** Seropédica, RJ, 2008. 72ip. Dissertation (Master's Degree in Animal Science). Federal Rural University of Rio de Janeiro.

OLIVEIRA, L.M.; PAIVA,R.; SANTOS, B.R.; PAIVA, P.D.O. Abiotic factors and plant production. In: PAIVA, R.; OLIVEIRA, L.M. (Ed.). **Plant physiology and production.** 1. ed. Lavras: Ed. UFLA, Brazil, 2006, p. 29-41.

ORTOLANI, A.A.; CAMARGO, M.B.P. Influences of climatic factors on production. In: CASTRO, P.R.C.; FERREIRA, S.O.; YAMADA. T. **Ecophysiology of agricultural production.**

Piracicaba: Ed. Brazilian Association for Potash and Phosphate Research. Brazil, 1987. p. 71-79.

PARSONS, A.J. The effects of season and management on the growth of grass swards. In: JONES, M.B.; LAZEMBY, A. (Ed.). **The grass crop: the physiological basis of production.** London: Ed.

Chapman and Hall, United Kingdon, 1988, p. 129-177.

PARSONS, A.J.; SCHWINNING, S.; CARRERE, P. Plant growth functions and possible spatial and temporal scaling errors in models of herbivory. **Grass & Forage Science,** v.56, n. l,p. 21-34, 2001.

PATÊS, N.M.S.; PIRES, A.J.V.; SILVA, C.C.F.; SANTOS, L.C.; CARVALHO, G.G.P.; FREIRE, M.A.L. Morphogenic and structural characteristics of tanzania grass submitted to doses of phosphorus and nitrogen. **Revista Brasileira de Zootecnia,** v.36, n.6, p.1736-1741, 2007.

PEDREIRA, C.A.G.; DA SILVA, S.C.; SOUZA NETO, J.M.; SBRISSIA, A.F. Grazing systems in Brazilian livestock farming. In: SYMPOSIUM ON STRATEGIC MANAGEMENT OF PASTURE, 2002, Viçosa. **Proceedings...** Viçosa, p. 197-234. 2002.

PEDREIRA, C.G.S.; MELLO, A.C.L.; OTANI, L. The forage production process in pastures. In: MATTOS, W.R.S. et al. (Ed.). **Animal production as seen by Brazilians.** Piracicaba, SP: Ed. FEALQ, Brazil, 2001, p. 772 - 807.

REIS, R.A.; MELO, G.M.P.; BERTIPAGLIA, L.M.A.; OLIVEIRA, A.P. *Cynodon* hay production. In: VILELA, D.; RESENDE, J.C.; LIMA, J (Ed.). ***Cynodon,* forage crops that are revolutionising Brazilian livestock farming.** 1. ed. Juiz de Fora: Ed. Embrapa Gado de Leite, Brazil, 2005, p.79-131.

RIBEIRO, R.C.; ROSSIELLO, R.O.P.; MACEDO, R.O.; BARBIERI JR, E. Introduction of desmodium in established Brachiaria humidicola pasture: density and frequency of the legume in the consortium. **Revista Universidade Rural, Série Ciências da Vida.** v. 27, n. 2, 2007. p. 41-49.

ROCHA, G. P.; EVANGELISTA, A. R.; LIMA, J. A.; ROSA, B. Nitrogen fertilisation in grasses of the genus *Cynodon.* **Ciência Animal Brasileira,** v. 3, n.l, p. 1-9, 2002

RODRIGUES, L.R.A.; RODRIGUES, T.J.D.; REIS, R.A.; SOARES FILHO, C.V. Evaluation of physiological characteristics of five *Cynodon* cultivars. **Acta Scientiarum Animal Science,** v.28 n.3, p.245-250, 2006a.

RODRIGUES, R.C.; ALVES, A.C.; BRENNECKE, K.; PLESE, L.P.M.; LUZ, P.H.C. Tiller density, dry mass production and leaf area of xaraés grass grown under doses of nitrogen and potassium.

Boletim de Indústria Animal.v.63, n.l, p.27- 33, 2006b.

RUSSELLE, M.P. Nitrogen cycling in pastures and ranges. **Journal of Production Agriculture.** v.5, p. 13-23,1992.

SARMENTO, P.; NASCIMENTO, R.C.; MARTINS, A.T.; CRUZ, M.C.P.; FERREIRA, M.E. Nutrients limiting the development of Tifton 85 grass in red-yellow clay. **Boletim de Indústria Animal,** v. 63, n.l, p.11-18, 2006.

SBRISSIA, A.F.; DA SILVA, S.C. Comparison of three methods for estimating leaf area index in marandu grass pastures under continuous stocking. **Revista Brasileira de Zootecnia,** v.37, n.2, p. 212-220, 2008.

SBRISSIA, A.F.; DA SILVA, S.C. The pasture ecosystem and animal production. In: MATTOS, W.R.S. et al. (Ed.). **Animal production in the view of Brazilians.** Piracicaba, SP: Ed. FEALQ, Brazil, 2001, p. 731 - 754.

SBRISSIA, A.F.; DA SILVA, S.C.; NASCIMENTO JÚNIOR, D. Ecophysiology of forage plants and grazing management. In: SYMPOSIUM ON PASTURE MANAGEMENT. 24, 2007, Piracicaba. **Proceedings of** the 24th Symposium on Pasture Management: Piracicaba, p. 1-27. 2007.

VENDRAMINI, J.M.B.; SILVEIRA, M.L.A.; DUBEUX JR., J.C.B.; SOLLENBERGER L.E. Environmental impacts and nutrient recycling on pastures grazed by cattle. **Revista Brasileira de Zootecnia,** v. 36, n. suplemento especial, p.139-149,2007.

VILELA, D. Potential of *Cynodon* pastures in dairy farming. In: VILELA, D.; RESENDE, J.C.; LIMA, J (Ed.). ***Cynodon,* forage crops that are revolutionising Brazilian livestock farming.** 1. ed. Juiz de Fora: Ed. Embrapa Gado de Leite, Brazil, 2005, p.165-176.

VILELA, D.; ALVIM, M. J. *Cynodon* pasture management: Introduction, characterisation and evolution of use in Brazil. In: SYMPOSIUM ON PASTURE MANAGEMENT, 15, Piracicaba, 1998. **Proceedings...** Piracicaba: FEALQ, p.23-54,1998.

XAVIER, D.F.; CARVALHO, M.M.; BOTREL, M.A.; FREITAS, V.P.; VERNEQUE, R.S. Effect of post-planting management on the establishment of elephant grass pasture. **Revista Brasileira de Zootecnia,** v. 30, n.4,\ p.1200-1203, 2001.

MIX
Papier aus verantwortungsvollen Quellen
Paper from responsible sources
FSC® C105338

Printed by Books on Demand GmbH, Norderstedt / Germany